DISEASES

James Shoals and
John Willis

LIGHTB◆X
openlightbox.com

LIGHTBOX

Go to
www.openlightbox.com
and enter this book's
unique code.

ACCESS CODE

LBXB4253

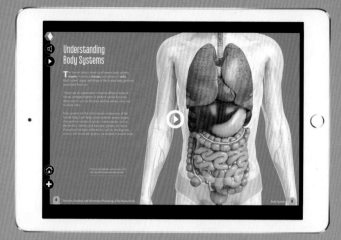

Lightbox is an all-inclusive digital solution for the teaching and learning of curriculum topics in an original, groundbreaking way. Lightbox is based on National Curriculum Standards.

STANDARD FEATURES OF LIGHTBOX

AUDIO High-quality narration using text-to-speech system

ACTIVITIES Printable PDFs that can be emailed and graded

SLIDESHOWS Pictorial overviews of key concepts

VIDEOS Embedded high-definition video clips

WEBLINKS Curated links to external, child-safe resources

TRANSPARENCIES Step-by-step layering of maps, diagrams, charts, and timelines

INTERACTIVE MAPS Interactive maps and aerial satellite imagery

QUIZZES Ten multiple choice questions that are automatically graded and emailed for teacher assessment

KEY WORDS Matching key concepts to their definitions

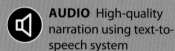

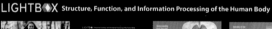

GROWTH, DEVELOPMENT, AND REPRODUCTION OF THE HUMAN BODY

DISEASES

Understanding Diseases and Disorders

Health is a state of the body in which the mind and the body both function properly. Disease, in contrast, refers to an **abnormal** condition affecting the body of an organism. It can affect the entire body or a few organ systems. The surroundings of human beings play a major role in the occurrence of diseases.

There are two basic types of diseases, infectious and noninfectious. Infectious diseases spread from one person to another. Diseases that do not spread from person to person are noninfectious. Many diseases are caused by microorganisms. These are small living things that cannot be seen with the naked eye. Various types of microorganisms cause diseases. They include bacteria, fungi, protozoa, and **parasites**. Viruses are tiny particles, or microbes. They also cause a range of diseases.

Disorders include a range of conditions that have a harmful effect on the body. All diseases are disorders, but other disorders include conditions that are not caused by disease. Examples of these types of disorders include broken bones, indigestion, and sunburn.

Researchers in laboratories study the tiny organisms that cause disease to learn how the body can resist them.

Classification of Diseases and Disorders

Diseases are classified in a number of different ways. One way is by how they are transmitted to sufferers. They can be transmitted by air or water, or by vectors, which are organisms or nonliving agents that carry the disease. Other diseases are caused by **genetic** problems or by **cells** within the body itself. Another way to classify diseases is by the main body systems they attack.

TRANSMISSION OF DISEASES

AIRBORNE DISEASE

Airborne diseases are transmitted by tiny bacteria in the air. We inhale them when we breathe.

WATERBORNE DISEASE

Waterborne diseases are transmitted in dirty water. They pass into the body when we drink.

VECTOR-BORNE DISEASE

Vector-borne diseases pass to humans from carriers, such as fleas, rats, or dogs.

TUMORS AND CANCERS

Tumors are body cells that form lumps. If the lumps are harmful, they are termed **cancerous**.

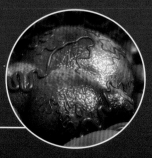

GENETIC DISORDERS

Genetic disorders are passed from parents to their children. Genes are units of **heredity**.

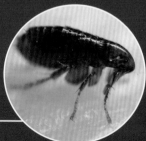

BODY SYSTEMS AFFECTED BY DISORDERS

CIRCULATORY SYSTEM

The circulatory system moves blood around the body to carry substances to and from the cells.

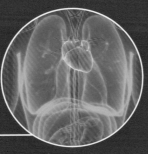

DIGESTIVE SYSTEM

The digestive system extracts nutrients from food to give the body energy.

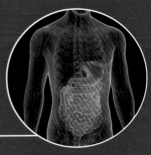

IMMUNE SYSTEM

The immune system defends the body against diseases.

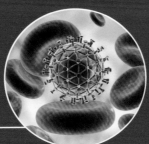

INTEGUMENTARY SYSTEM

The integumentary system includes the skin, hair, and nails. It protects us from things outside the body.

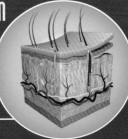

MUSCULOSKELETAL SYSTEM

The musculoskeletal system supports the human body and allows it to move.

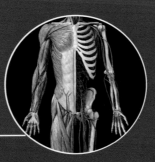

NERVOUS SYSTEM

The nervous system carries messages between the body and the brain.

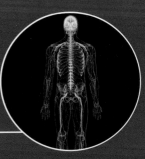

RESPIRATORY SYSTEM

The respiratory system controls the inhalation and exhalation of air in the body.

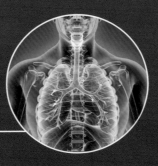

SENSORY SYSTEM

The sensory system transmits information to the brain from the sense organs.

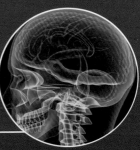

URINARY SYSTEM

The urinary system removes unwanted, undigested waste from the body.

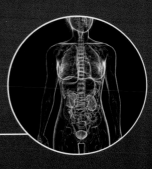

Timeline of Fighting Infectious Diseases

The fight against disease began seriously in the mid-1860s. Biologists discovered that diseases were spread by germs. This discovery enabled a series of discoveries that helped prevent and cure diseases. However, some diseases remain a serious problem today.

1865
The surgeon Joseph Lister begins washing his hands with carbolic soap before performing operations.

1700	1750	1800	1850	1900

1749
The English doctor Edward Jenner introduces the first **vaccine** to be used against smallpox.

1847
Austrian doctor Ignaz Semmelweis recommends that surgeons wash their hands before delivering babies. This dramatically reduces the number of illnesses in newborn babies.

1861
French biologist Louis Pasteur helps spread the idea that diseases are caused by bacteria. This is now called the germ theory of disease.

1971
A single vaccine is introduced to prevent people from catching measles, mumps, and rubella. It is called the MMR vaccine.

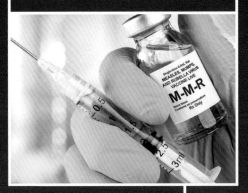

1979
Smallpox, which was once a major killer, is **eradicated** worldwide.

1925	1950	1975	2000	2020

1928
Alexander Fleming discovers penicillin. The drug can kill germs and prevent infections.

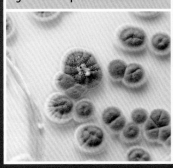

2020
COVID-19, a respiratory disease, rapidly spreads worldwide. In March, the World Health Organization (WHO) declares the outbreak a pandemic.

Causes of Disease

There are many causes of disease. They are all transmitted in a limited number of ways, however. The way a disease spreads is important to scientists who study diseases. These scientists are called epidemiologists. Studying the spread of different diseases helps them develop methods to prevent a particular disease from spreading.

Epidemiologists study possible vectors of disease, such as infected birds and other animals.

Disease is the **number 1** cause of death in the world.

Nearly **1.3 million** people worldwide died of the disease tuberculosis in 2016.

The U.S. Centers for Disease Prevention and Control has high levels of concern about **14** vector-borne diseases.

Airborne Diseases

Disease-causing agents float around in the air. Diseases that can be passed from one person to another through the air are known as airborne diseases. Tiny pathogens are generated while coughing, sneezing, or talking. Diseases can also spread due to contact with saliva or nasal discharges, and from close mouth-to-mouth contact.

Vector-borne Disease

Vector-borne diseases are diseases that are spread through vectors such as bacteria, fungi, viruses, protozoa, and parasites. Flies, ticks, bats, rats, and dogs can also act as vectors

Hairs and mucus in our noses help prevent airborne diseases from entering the body.

Drinking dirty or stagnant water runs a high risk of causing waterborne disease.

for the transmission of different diseases. **Precautionary** measures against vector-borne diseases include having vaccinations, maintaining good hygiene, and using mosquito repellents and nets to prevent insect bites.

Waterborne Disease

Waterborne diseases spread through water that is contaminated with microorganisms. The contamination of water can occur in many ways. These include accidental mixing of harmful substances in water, **fecal** contamination of water, and improper cleaning. Most waterborne diseases are not fatal, but they can be extremely unpleasant.

Disorders of the Circulatory System

The heart and blood are part of the circulatory system. The heart pumps oxygen-rich blood from the lungs to the rest of the body. Blood is comprised of red blood cells, white blood cells, platelets, and a liquid called plasma. All these parts of blood perform different functions, such as carrying oxygen to different body parts and fighting infections that enter the body. Heart disorders are potentially serious because they halt or obstruct the circulation of blood. Blood disorders, meanwhile, can affect the whole body.

Anemia

Anemia is a common blood disorder that involves a lack of red blood cells, which carry oxygen around the body. Because oxygen is less available in the body, someone suffering from anemia feels pale and weak, gets headaches, and has difficulty concentrating. Eating foods rich in iron and vitamins can help avoid anemia.

The body continually makes new blood, so people can donate blood to others with blood disorders.

BLOOD TESTS

Blood tests are very useful to doctors. The substances in blood change when a person is suffering from a disease or disorder. Blood tests reveal if organs are not working well, for example. The tests are carried out by health workers called phlebotomists.

Blood tests require only a little blood and are not painful.

Blood pressure is measured using a tight cuff that detects blood movement in the upper arm.

Chest pains can be a sign of serious disorders, so always consult a doctor if they occur.

Angina

Angina is a pain in the chest. It arises when the **arteries** of the heart are blocked. This reduces the flow of blood and fresh oxygen to the heart.

Blood Pressure

Blood presses on the walls of blood vessels as it moves around the body. High or low blood pressure can cause medical problems. High blood pressure is also called hypertension. Its symptoms include shortness of breath, headaches, and chest pain. Eating a healthy, low-fat diet and exercising can reduce blood pressure. Low blood pressure is called hypotension. Its symptoms include lack of concentration, blurred vision, tiredness, and depression.

Blood pressure can be raised by eating regular small meals and drinking more water.

Heart Attack

A heart attack is caused by a sudden obstruction in the blood flow to the heart. The symptoms of a heart attack include prolonged chest pain, pressure in the chest, heavy sweating, **palpitations**, and fainting. A doctor should be consulted immediately.

Heart Failure

Heart failure is a disorder in which the heart is unable to supply enough blood to the body parts. Indications of heart failure include swollen legs, fatigue, angina, loss of appetite, and abnormal weight loss or gain.

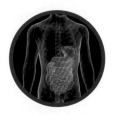

Disorders of the Digestive System

The digestive system is made up of several organs, such as the stomach, **pancreas**, liver, and intestines. The system converts the food that we eat into useful energy for the body. The digestive system is affected by several common diseases and disorders.

Constipation

Constipation is a common intestinal problem. It is characterized by hard and dry **stools**, or an inability to go to the bathroom. The condition is caused by excessive absorption of water from the intestine. Symptoms of constipation include **abdominal** cramps and pain. Frequently drinking water and eating a fiber-rich diet help ease the problem. Medication such as laxatives also relieves the symptoms of constipation.

Gallstones

Gallstones are hard deposits formed by digestive juices in the gall bladder. Gallstones can cause stomach pain, nausea, and vomiting.

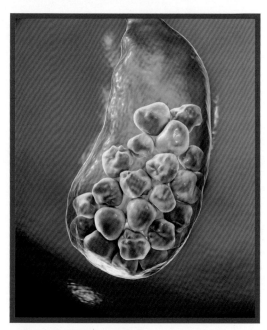

Gallstones become painful when they move from the gall bladder into the small intestine.

CURRICULUM CORRELATION

Why do bacteria thrive in the stomach?

The bacteria that can cause stomach ulcers are single-celled organisms. Like all other organisms, they are encouraged to reproduce and grow by their environment. Research the factors that make the stomach a good environment for bacteria to thrive.

Heartburn

Heartburn affects the stomach and the food pipe. It is caused by the upward flow of acid produced in the stomach. This acid sometimes enters the food pipe and irritates the lining, which causes a burning sensation and chest pain. Avoiding hot and spicy foods is a simple step that can help in controlling heartburn.

Indigestion

Indigestion is also called dyspepsia. It can be caused by minor problems, such as overeating. It also has more serious causes, such as stomach ulcers and cancer. Symptoms of dyspepsia include a constant feeling of being full, nausea, vomiting, and abdominal pain. Mild indigestion can be avoided by properly chewing food, eating low-fat meals, and avoiding coffee.

Lactose Intolerance

Lactose intolerance is the body's inability to digest milk and milk products. It is a very common disorder. Lactose intolerance arises when the body produces insufficient amounts of lactase, which is an **enzyme** that breaks down lactose, the sugar in milk. The undigested milk may cause stomach pain, gas, and diarrhea. The best treatment is simply to avoid dairy products.

Despite its name, heartburn has nothing to do with the heart apart from occurring nearby in the body.

Stomach Ulcer

Stomach ulcers, or gastric ulcers, are sores on the stomach lining. They can be caused by bacterial infection or by a rise in the amount of acids in the stomach. Common symptoms of stomach ulcers are changes in the color of the stools, stomach pain, nausea, and heartburn. Ulcers need medication to heal.

About 65 percent of the world's population has some level of lactose intolerance.

Disorders of the Eye and Ear

The sense organs are the eyes, ears, nose, tongue, and skin. They help us to perceive the world around us. The eyes and ears are particularly sensitive. They are easily affected by different diseases and disorders. The most extreme conditions are blindness and deafness. There are many other less serious disorders of both the eyes and the ears.

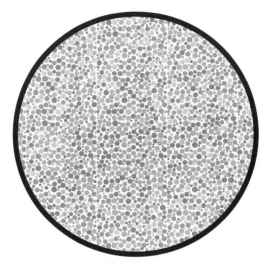

The most common form of color blindness is an inability to tell the colors green and red apart.

Cataract

A cataract is a condition in which the eye lens becomes clouded due to **protein** deposits. It leads to hampered vision and reduced color visibility. Cataracts may lead to blindness. Surgery is the most effective treatment for cataracts.

Color Blindness

Color blindness is an incurable disorder of the eye. A color-blind individual cannot see some or all the colors that most individuals can see. Color blindness arises due to the absence of color granules in the cone cells that detect light at the back of the eye.

Ear Infection

Otitis media, or ear infection, is a type of infection of the middle ear. The infection is characterized by the presence of fluid and mucus in the ear's tubes. The condition is painful and reduces a person's ability to hear. This type of infection is quite common in young children. It is not usually serious, but it should be addressed with a doctor to ensure that there is no permanent damage.

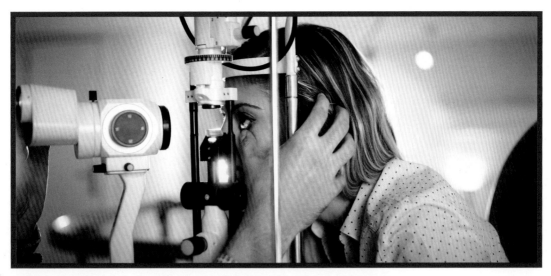

Optometrists use a range of tests to diagnose farsightedness and nearsightedness.

Farsightedness and Nearsightedness

Farsightedness and nearsightedness are common vision problems. People who are farsighted can see things well at a distance but not close up. Nearsighted people are the reverse. In both cases, the eye does not focus light on the retina. The conditions can be managed by wearing eye glasses or contact lenses. They may also be corrected through laser surgery.

Pink Eye

Pink eye, or conjunctivitis, is a common communicable eye disease that causes itching and burning. The causes of pink eye are bacterial infection, **allergies**, and irritants such as gaseous fumes, eye-care products, and contact lenses.

Sty

A sty is a bacterial infection of the eyelids. The infection blocks the secretionary **glands** on the eyelids. The eyelid becomes swollen and painful. A sty can be caused by stress, overuse of cosmetics, or poor hygiene.

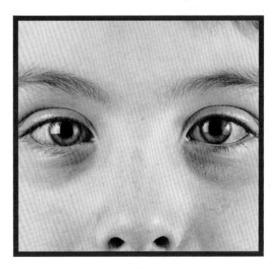

Pink eye occurs when tiny blood vessels burst in the white part of the eye.

Genetic Disorders

Genes are how parents pass on characteristics to their children. Genetic disorders occur when genes are missing or faulty. Some genetic diseases can be life threatening, but many are manageable.

Hemophilia

Hemophilia is a genetic disorder in which blood does not **clot** normally. The disease is caused by a faulty gene that fails to produce a protein that makes the blood coagulate. Although the disease is rare, it can be fatal if it leads to excessive blood loss or blood loss in the brain. Hemophilia ranges from mild to severe. Treatment involves the injection of a substance that helps the blood to clot.

Genes are carried on long, twisting molecules of deoxyribonucleic acid (DNA).

Muscular Dystrophy

Muscular dystrophy (MD) is a genetic disorder that affects the muscles. Missing or damaged genes prevent individuals from producing proteins that keep the muscles healthy. MD is a **degenerative** disease that can complicate simple activities such as walking. It is more common in boys than girls. There is no cure, but treatment can slow the progression of the disease.

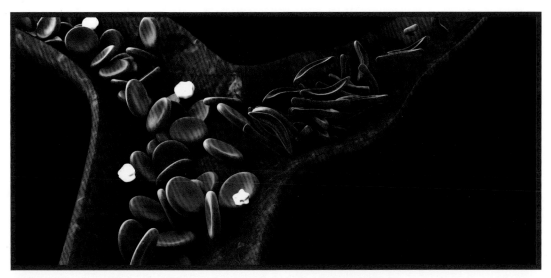

Sickle cells clump together at junctions in narrow blood vessels, causing a blockage.

Sickle Cell Anemia

Sickle cell anemia causes blood cells that are usually disk-shaped to become crescent-shaped. The misshapen cells are unable to transport enough oxygen to the body. They can also clog blood vessels, which obstructs normal blood flow. The symptoms of sickle cell anemia include attacks of severe pain, fatigue, frequent infections, strokes, and delayed growth. Patients require treatment including blood **transfusions** and painkillers.

Thalassemia

Thalassemia is a blood-related genetic disorder. In this disorder, the body does not produce hemoglobin, which is a protein that is normally present in the blood. A lack of hemoglobin limits the blood's ability to carry oxygen around the body. People suffering from thalassemia experience mild to severe anemia, fatigue, bone deformities, abnormal growth, and **jaundice**. The disease is usually managed using medicines and blood transfusions.

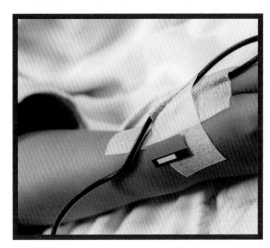

People with the most severe form of thalassemia require blood transfusions about once a month.

Disorders of the Immune System

The immune system is the defensive system that helps the body to fight off infections. It is made up of many different types of cells, tissues, and organs. It is frequently attacked by diseases.

Diabetes Mellitus

Type 1 diabetes mellitus is an inflammatory **autoimmune** disease of the pancreas. A person suffering from diabetes is not able to produce enough insulin in the body. This **hormone** helps regulate carbohydrate and fat processing. This form of diabetes is generally controlled by administering insulin.

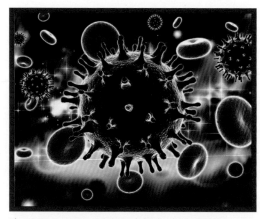

The HIV virus moves around the body in the bloodstream, where it infects white blood cells.

Graves' Disease

An excess release of the thyroid hormone from the thyroid gland leads to Graves' disease. A higher production of the thyroid hormone causes eye and vision problems, heat intolerance, goiter, insomnia, and increased appetite.

HIV/AIDS

Acquired Immunodeficiency Syndrome (AIDS) is a viral disease. It is caused by the HIV virus, which weakens the immune system. This makes infected individuals susceptible to other diseases. AIDS is the final stage of HIV infection. HIV is called a retrovirus. It multiplies by mixing with the host cell's DNA.

Pernicious Anemia

Pernicious anemia is an autoimmune disorder. It occurs when there is very low absorption of vitamin B12 by the digestive tract. Vitamin B12 is necessary for the proper development and functioning of the red blood cells. Lack of this vitamin leads to low levels of red blood cells in the body.

This may lead to lack of energy, loss of appetite, pale skin, or numbness of the hands and feet.

Primary Immunodeficiency

Primary immunodeficiency disorders occur when a part of the immune system is either missing or does not function properly. Such disorders may occur due to excessive use of medicines and drugs, other diseases, and exposure to chemicals. These diseases weaken the immune system and make it prone to infections.

Rheumatic Fever

Rheumatic fever is an autoimmune inflammatory disease. It is caused by bacteria and leads to the swelling of the heart valves, brain, skin, and **joints**. The disease commonly affects

Sufferers with rheumatic fever have a temperature of over 100° Fahrenheit (38° Celsius).

children between 6 and 15 years of age. The disease is characterized by the presence of fever, abdominal pain, heart problems, skin problems, and muscle weakness. Rheumatic fever is a serious disease, but it can be treated with antibiotics.

Scleroderma

Scleroderma is an autoimmune disorder that leads to deposits of protein under the skin. The affected person experiences thickening and hardening of the skin of the hands and face, stiffness of the body, and joint pain. The symptoms can be reduced by using medication.

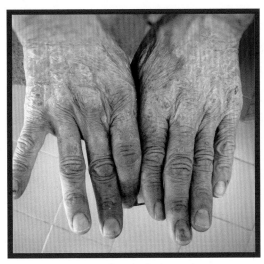

Scleroderma occurs when the body produces too much of a protein called collagen.

Disorders of the Integumentary System

The integumentary system is the outer layer of the body. It includes the skin, hair, and nails. The skin is the organ of touch and sensation. It is made up of three layers, each of which performs a different function. The skin helps us to respond to our surroundings. Since the skin is the outer covering of the body, it can be affected by many disease-causing agents.

Acne

Acne is a common skin problem. It usually emerges during **adolescence**. It occurs in the form of small whiteheads and blackheads on the face, neck, and chest region. Hormonal changes in the body are a main cause of acne. Maintaining good personal hygiene helps reduce the problem of acne.

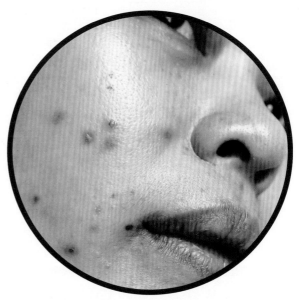

In serious cases of acne, people can take medications, such as antibiotics.

Cancer

Cancers are an unrestrained growth of cells. There are a number of different types of skin cancer, some more serious than others. The most serious type is melanoma.

CURRICULUM CORRELATION

Will environmental change make skin disorders more serious?

Dangerous ultraviolet rays in sunlight damage the skin. Human activity has changed the environment by reducing the amount of ozone in the atmosphere. Research other ways changing the environment like this has a negative impact on humans.

Dandruff

Dandruff is a skin problem that affects the scalp, the skin covering the head. It is formed by a thin layer of dead skin cells pushed off the scalp. Dandruff is normal, but a high amount of dandruff can cause discomfort and lead to skin allergies.

Regular use of anti-dandruff shampoo can help reduce the effects of dandruff.

Erythema

Erythema is a redness in the skin. It can arise due to an injury, underlying skin disease, acne medication, allergies, waxing, and tweezing of the hairs. Erythema can occur on any part of the body. A doctor should be consulted for proper treatment.

Sunburn

A sunburn is burning of the skin due to overexposure to the Sun. The Sun's rays are made up of different types of **ultraviolet** (UV) rays that are harmful for the skin. As a result of the sunburn, the skin turns red. In severe cases, the person may feel dizziness and fatigue. Sunburns can be treated using medicines.

The best way to avoid sunburns is to prevent them entirely by using sunscreen and wearing a hat.

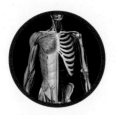

Disorders of the Musculoskeletal System

Bones, muscles, and joints are part of the musculoskeletal system. The skeletal system forms the supporting framework for the body and gives a fixed shape and form to the body. Muscles stretch and tighten to move the bones when people move. Bones, muscles, and joints are all susceptible to diseases and disorders, such as breaks.

Bone Fracture

A bone fracture is an injury to a bone that causes cracking or breakage. Fractures can occur for several reasons, such as bone diseases, falls, or accidents. Different types of fractures are treated in different ways. Usually, a cast is applied to keep the injured bone in place as it heals.

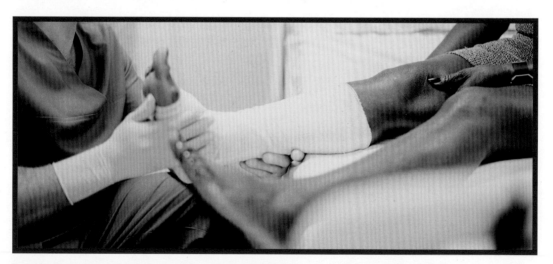

Bone fractures can occur in any part of the body, such as the legs, arms, hips, backbone, or ribs.

CURRICULUM CORRELATION

How can gene therapy help broken bones?

In some cases, broken bones can be encouraged to heal by inserting genes that promote faster bone growth. Research how gene therapy like this can have beneficial effects for a person.

Clubfoot

Clubfoot is a birth defect of the muscle and bone tissue of the feet. This condition causes the foot to be turned inward and downward. The muscles of the affected foot are shorter than normal. The foot may be flexible or rigid and has reduced movement. Foot-reshaping treatment can be used to resolve the disorder.

Gout

Gout is a form of **arthritis** that affects the joints of the body. The problem occurs due to excess uric acid deposits in the blood. People suffering from diabetes, kidney diseases, obesity, and blood cancers have a higher risk of developing gout. The problem usually begins from the toe, knee, or ankle joints and leads to painful, swollen, and stiff joints.

Osteoarthritis

Osteoarthritis is a degenerative disease of the joints. Its main cause is the destruction of **cartilage**. Cartilage is present around the joints and helps in smooth movement of the bones. The destruction of the cartilage causes friction when bones move against each other. This condition leads to the wear and tear of joints, causing persistent pain, swelling, and stiffness.

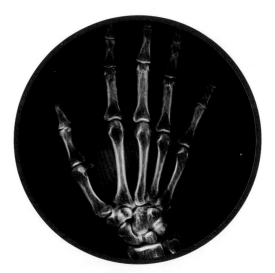

Inflammation is the immune system's response to what it mistakes for a harmful substance in the joints.

Rheumatoid Arthritis

Rheumatoid arthritis is an autoimmune disease of the joints. The immune system of the body attacks healthy joints and leads to their inflammation. The disease commonly affects the wrists, fingers, knees, feet, and ankle joints. The joints become tender, develop pain, and become deformed with time.

Sports Injury

Sports injuries occur while exercising or playing games. These injuries may affect any part of the bones, muscles, or joints. They include sprains and strains, knee injuries, muscle pains, bone fractures, and bone dislocations. Treatment of sports injuries includes proper rest, use of pain relievers, and surgery.

Disorders of the Nervous System

The nervous system is the body's control-and-coordination system. It comprises the brain, spinal cord, and many different types of **nerves** that are spread throughout the body. Any type of disorder or disease in the nervous system can affect any part of the body. Nervous system diseases can occur because of a trauma, developmental disorders, and genetic errors. They are degenerative in nature.

A trauma such as banging one's head in a fall can damage the nervous system.

Alzheimer's Disease

Alzheimer's disease is a degenerative disease of the brain and nervous system. It usually occurs in older people. The disease progresses slowly. It leads to a gradual loss of memory and thinking ability.

Epilepsy

Epilepsy is a brain disorder characterized by regular seizures. A seizure is a small period during which brain activity reduces and the brain does not work properly. Several causes may lead to epilepsy. They include brain injury, heart attack, birth defects, and other disorders.

The average age for the onset of Alzheimer's disease is **65 YEARS**.

Up to **20 percent** of strokes are fatal. Many more cause long-term health damage.

Diseases of the nervous system cause about **100 million** people to suffer serious pain.

Slipped Disk

A slipped disk is a disorder of the lower backbone. It arises when a vertebral disk present between the bones of the spinal cord dislocates from its normal place. This causes severe back pain. A person suffering from a slipped disk may also experience numbness, a burning sensation, and weakness.

Spondylosis

Spondylosis is a common degenerative disease of the backbone. It is caused by the normal wear and tear of the joints and tissues. The disease can affect the upper, middle, or lower spine. The occurrence of spondylosis is marked by a loss of sensation in the arms, legs, and shoulders, sudden numbness, and loss of balance.

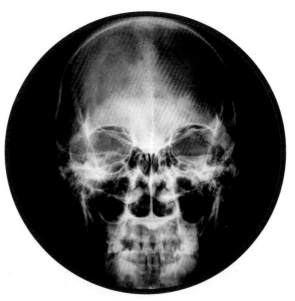

Strokes can lead to long-term conditions such as poor speech. In some cases, they are fatal.

Stroke

A stroke is a condition of the brain in which a clot stops blood flow. It causes numbness in different parts of the body, loss of coordination, and improper body functioning.

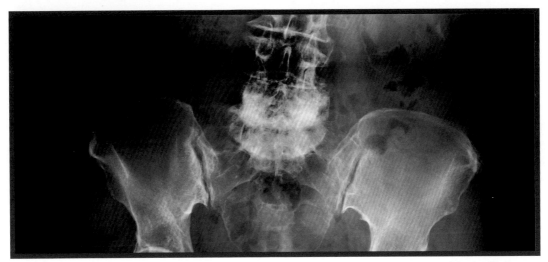

Spondylosis can be caused by the degeneration of intervertebral disks between the vertebrae of the backbone.

Disorders of the Respiratory System

The respiratory system includes the nose, trachea, **bronchi**, chest cavity, lungs, and **diaphragm**. Its main function is to provide the body with oxygen. The respiratory system is particularly vulnerable to diseases and disorders. The air includes many dust particles and microorganisms. When inhaled, these components can harm various parts of the respiratory system. This can lead to a range of infections and diseases.

Asthma

Asthma is a respiratory disease of the air tubes, or bronchi. Whenever any foreign particle enters the air tubes, the tubes swell up, reducing the air flow to the lungs. This may lead to severe coughing, suffocation, wheezing, and breathing problems. Asthma is fairly common in children, but some outgrow it as adults. People with asthma usually try to avoid situations that trigger attacks.

RHINOVIRUSES

Rhinoviruses are the most common viruses that cause illness among humans. "Rhino" means nose, and the viruses live in the nose. Rhinoviruses are the cause of the common cold, among other infections. The virus is transmitted easily on droplets in sneezes and coughs, or passed on via hands or surfaces.

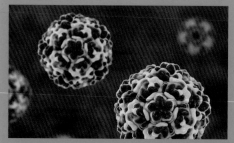

Rhinoviruses thrive in the warm, damp conditions inside the nose.

There is no cure for the common cold. The best treatment is to rest and to drink fluids.

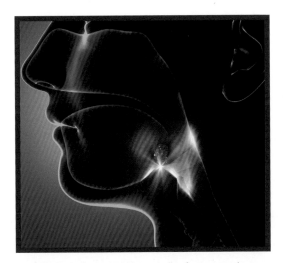

Tonsillitis can be caused by a viral infection, such as the common cold, or by a bacterial infection, such as strep throat.

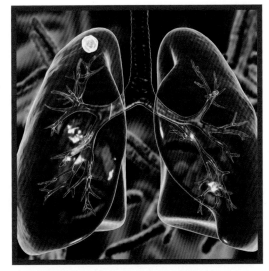

Tuberculosis most often affects the lungs, but can also affect the glands, bones, and nervous system.

Bronchitis

Bronchitis is an inflammation of the mucous membrane in the bronchi. It takes two forms, **acute** and **chronic**. The acute form is caused by bacterial infection and tends to follow other respiratory infections. It usually results in a serious cough and difficulty breathing.

Common Cold

The common cold is one of the most familiar types of respiratory infections. It occurs due to viral infection in the upper respiratory tract. People with a cold frequently experience symptoms that include an increase in body temperature, headache, cough, sore throat, sneezing, and a runny nose. Most colds pass within a few days.

Tonsillitis

The tonsils are tissues in the throat that filter bacteria and viruses breathed in through the mouth and the nose. Tonsillitis is caused when the tonsils become inflamed. It causes severe throat pain, problems in swallowing, fever, and bad breath.

Tuberculosis

Tuberculosis (TB) is an infectious disease of the respiratory system caused by bacteria. The infection begins in the lungs but can spread to other parts of the body. A persistent cough, weakness, fever, chills, and weight loss are symptoms associated with TB. The disease is curable with proper medical intervention, which might include numerous treatments.

Tumors and Cancer

A tumor is an abnormal condition in which the cells of a particular part of the body keep on growing and dividing. The tumors may or may not be visible externally. Tumors are either benign or malignant.

Benign tumors are restricted to one part of the body and do not spread. They either do not grow or only grow slowly. Malignant tumors are also called cancerous tumors. They grow very quickly and can spread to other parts of the body. They can cause serious illness, and in many cases may lead to death.

The best chance of curing cancer is for it to be detected early.

Brain Tumors

Brain tumors are tumors that affect any part of the brain. When a tumor begins in the brain, it is called a primary brain tumor. When tumors that originate elsewhere spread to the brain, they are called secondary brain tumors. Brain tumors produce recognizable symptoms. These include severe headaches, seizures, changes in mental ability, changes in the senses, and a sudden inability to read or write.

Leukemia

Leukemia is a type of blood cancer. This condition produces abnormal white blood cells in the bone marrow. A person suffering from leukemia is prone to several types of infections, anemia, and weight loss due to weak white blood cells.

Lung Cancer

Lung cancer is a common type of cancer of the respiratory system. It may affect any part of the respiratory system, including the air tubes and lungs. Smoking cigarettes is the leading cause of lung cancer, but it can also be caused by exposure to certain pollutants. Common symptoms include bloody coughs, chest pain, recurring **pneumonia**, loss of appetite, and fatigue.

Oral Cancer

Oral cancer can affect any part of the mouth, such as the lips, tongue, gums, and palate. The cancer usually begins as a dark-colored mouth ulcer or as white plaque, and gradually spreads. The condition may become very painful, causing difficulties in speaking and swallowing food.

Skin Cancer

Skin cancer affects the skin cells. It usually appears on the head, neck, face, and arms. The most common causes of skin cancer are repeated or prolonged sunburns and a family history of skin cancer. Some skin cancers begin at the cells of the outermost skin layer, the epidermis. One serious type of skin cancer is called melanoma. It begins in the melanocytes, which are the cells that give skin its color. The first appearance of skin cancer usually takes the form of a nonhealing sore on the skin.

Treatments for cancer often cause other changes in the body, such as the loss of hair.

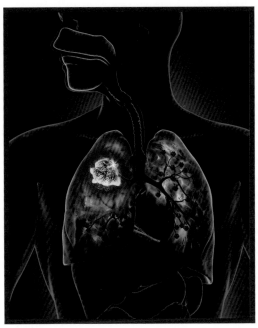

In lung cancer, cells divide uncontrollably in the lungs and cause tumors that restrict breathing.

Disorders of the Urinary System

The urinary system works continuously to filter the blood and extract waste products from it. This process begins in the kidneys. The waste products are further filtered, and important nutrients and water are absorbed back into the body. The liquid waste is called urine. It is ejected from the body via the bladder and a tube called the urethra. The urinary system is also affected by several diseases and disorders. Many of these conditions are minor, but some are more serious.

One sign of a urinary disorder is having to go to the bathroom frequently during the night.

Hematuria

Hematuria is a medical condition in which a large number of red blood cells are passed in the urine. The condition can often be recognized due to the red color of the urine. Hematuria may occur due to extremely high levels of stress or trauma. It can also be brought on by urinary tract infections (UTIs), kidney stones, and some types of cancers. Anyone finding blood in their urine should consult with a doctor immediately.

Kidney Failure

Kidney failure is a serious condition in which the kidneys do not function properly. This causes the amount of wastes in the body to increase. The condition may affect one or both kidneys. It may lead to several other problems, such as poisoning of the body, an increase in blood pressure, anemia, delayed healing from injury, and the inhibition of body functions. Kidney failures can be treated with kidney transplants or dialysis.

Kidney Stones

Kidney stones are **crystalline** deposits in the kidneys. One cause is prolonged **dehydration**. The lack of water causes urine to become more concentrated. Obesity can also contribute to kidney stones, as can certain types of diets. Kidney stones cause severe pain, swelling of the kidneys, and UTIs. In most cases, the stones move out of the body on their own, but in some cases surgical removal is required.

A kidney dialysis machine takes over the role of the kidneys. It filters a person's blood artificially.

Urinary Tract Infection

A UTI can occur in any part of the urinary system. The most common type of UTI is urethritis, which affects the urethra. A person suffering from a UTI will likely experience pain or a burning sensation during urination. Other symptoms include fever, cloudy or foul-smelling urine, fatigue, and lower abdominal pain. UTIs are unpleasant but treatable.

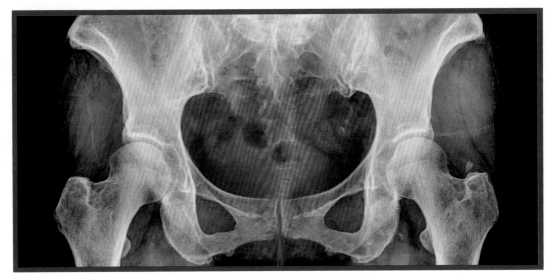

Cystitis is a type of infection that causes a burning sensation in the lower urinary tract.

Infectious Diseases

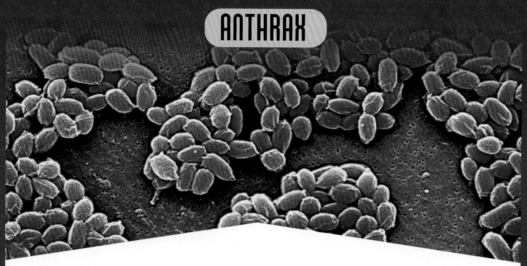

ANTHRAX

The spores of the anthrax bacteria are present in the air. They enter the body when inhaled or via open cuts or sores. Anthrax bacteria can also infect human beings through contact with the skin and hairs of animals such as sheep and goats. The bacterial infection usually leads to skin and lung diseases.

Means of transmission: Airborne **Type of disease:** Bacterial
Body systems affected: Integumentary system, respiratory system **Prevention:** Vaccination

CHICKEN POX

Chicken pox (varicella) spreads through sneezing and coughing of infected individuals. It can also spread via physical contact with the rashes. The infection leads to the formation of red, fluid-filled blisters on the skin. The viral infection cannot be treated with antibiotics but tends to clear up on its own in a few days.

Means of transmission: Airborne **Type of disease:** Viral
Body systems affected: Integumentary system **Prevention:** Vaccination

CHOLERA

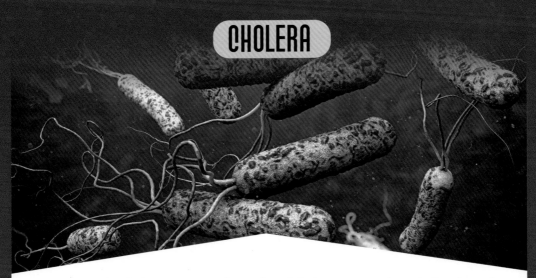

Cholera is spread by drinking contaminated water. Bacteria in the water cause the production of excessive water in the small intestine. The symptoms are severe pain in the abdomen, watery diarrhea, and abnormal thirst. Cholera often occurs in places that have poor hygiene and during times of flood or famine.

Means of transmission: Waterborne **Type of disease**: Bacterial
Body systems affected: Digestive system **Prevention:** Personal hygiene, environmental hygiene

DYSENTERY

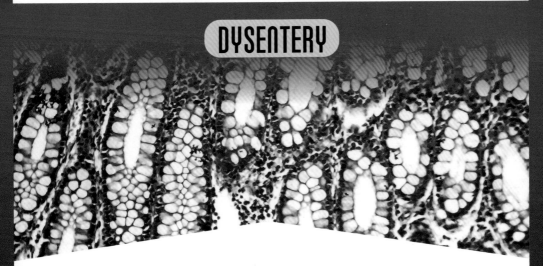

Dysentery spreads through the consumption of contaminated food and water. It affects a part of the intestine, causing severe diarrhea. The infection can be caused by bacteria or intestinal parasites. The disease can be cured with medication and intravenous fluids.

Means of transmission: Waterborne **Type of disease:** Bacterial, parasitic
Body systems affected: Digestive system **Prevention:** Personal hygiene

INFLUENZA

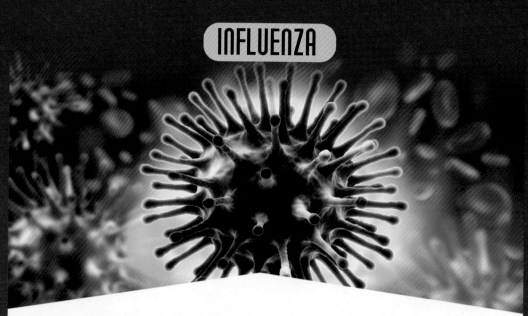

Influenza, or the flu, is caused by many different types of viruses. The symptoms of the disease include a sore throat or cough, headaches, chills, and aches in the body. The disease can be deadly for newborn babies, elderly people, and people with a weak immune system.

Means of transmission: Airborne **Type of disease:** Viral
Body systems affected: Respiratory system **Prevention:** Personal hygiene, vaccination

MALARIA

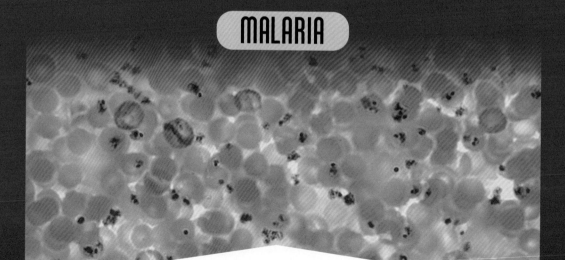

Malaria spreads through the bite of the female *Anopheles* mosquito. The insect bite transfers the parasites to the bloodstream. In addition to fever, the infection causes severe chills, vomiting, jaundice, and diarrhea. Malaria is a very serious illness, but it can be treated with proper medication.

Means of transmission: Vector-borne **Type of disease:** Parasitic **Body systems affected:** Circulatory system, nervous system, integumentary system, digestive system **Prevention:** Vaccination, hygiene

MEASLES

Measles (rubella) is highly contagious. It spreads through the respiratory droplets of an infected person. The disease usually begins with a high fever, cough, and runny nose. Red rashes develop after a few days. Measles is extremely dangerous to pregnant women because it can hurt or kill their developing babies.

Means of transmission: Airborne **Type of disease:** Viral
Body systems affected: Respiratory system, integumentary system **Prevention:** Vaccination

YELLOW FEVER

Yellow fever is spread through the *Aedes* mosquito in tropical areas. The disease is so named due to one of the symptoms, jaundice, which turns the skin yellow. The infection also causes high fever and chills, nausea, muscle pains, headache, and black vomiting. There is no known cure for the disease.

Means of transmission: Waterborne **Type of disease:** Bacterial, parasitic
Body systems affected: Digestive system **Prevention:** Vaccination, insect repellent

The World of Infectious Diseases

Curing diseases is a global effort. Doctors and scientists around the world are also attempting to find better ways to stop diseases from occurring. When diseases do occur, health experts lead efforts to prevent them from spreading.

Arctic Ocean

NORTH AMERICA

Pacific Ocean

Atlantic Ocean

SOUTH AMERICA

Baltimore, United States
Johns Hopkins Hospital is one of the leading centers for fighting diseases around the world. It is frequently rated as the best hospital in the United States. As well as treating patients in Baltimore, the hospital offers remote consultations to patients worldwide.

MAP LEGEND
- ☐ Land
- ☐ Water

N

SCALE
0   1,000 miles

1,000 kilometers

London, United Kingdom
The London School for Hygiene and Tropical Medicine is a public research university. It is a world authority on conditions that thrive in tropical climates, such as malaria.

Arctic Ocean

Stockholm, Sweden
The European Centre for Disease Prevention and Control (ECDPC) is based in Stockholm. It leads the European Union's fight against diseases. The ECDPC coordinates research and advises governments on threats from infectious diseases.

EUROPE

ASIA

Pacific Ocean

AFRICA

Indian Ocean

AUSTRALIA

Southern Ocean

Bangkok, Thailand
The Bumrungrad International Hospital treats more than 1.1 million patients each year. The hospital is noted for its pioneering use of medical technology to diagnose diseases in patients.

Prevention of Infectious Diseases

Disease prevention is necessary in order to avoid the large-scale spread of diseases and for the overall social and economic well-being of society. The prevention of disease takes place on different levels. It includes individual decisions as well as action taken at a community or government level.

PERSONAL HYGIENE

Personal hygiene refers to hygiene measures that are followed by individuals. Personal hygiene measures include practices such as bathing, keeping the ears and nose clean, washing clothes regularly, and washing hands before and after eating food or using the bathroom.

CURRICULUM CORRELATION

What is the impact of environment on disease?

The microorganisms that cause disease are like all other organisms. They prefer certain conditions that help them survive and multiply. Research some of the most common environmental conditions that encourage the spread of disease.

LIFESTYLE CHANGES

Basic changes in lifestyle can help in maintaining a healthy body and reducing the risk of diseases. Getting more sleep, reducing junk food in favor of a balanced diet, and increasing physical activity are all ways to improve health and strengthen the immune system.

COMMUNITY HEALTH

Community health practices include the measures and services provided by the government in order to maintain good health. Governments can take steps to create awareness about diseases and provide health-care services, such as free or low-cost vaccinations. In addition, governments take responsibility for efficient sewage systems, garbage disposal, and the provision of clean and safe drinking water.

ENVIRONMENTAL HYGIENE

Environmental hygiene refers to the maintenance of clean and healthy living conditions in and around human habitats. It includes practices such as trash removal and regular cleaning of homes and streets. One important hygiene issue in warm areas is stagnant water. Stagnant water offers an ideal breeding ground for mosquitoes, which carry a wide range of infectious diseases.

Health Care Careers

There are many health care careers related to diseases and disorders. There are jobs for research scientists who try to learn ways to prevent or cure disease. There are also a range of jobs for people who want to directly help those suffering from diseases by offering specialized or more general care.

SONOGRAPHER

A sonographer, or ultrasound technologist, uses sound waves to take images of the body's internal organs and blood vessels. This helps physicians diagnose diseases and disorders. Most ultrasound machines are in hospitals and clinics. Ultrasound technologists often work in dimly lit environments, where the technology functions best.

Duties: Determining what ultrasound images to take, and taking them

Education: A degree or an associate degree in ultrasound technology

Interests: Attention to detail, using technology

SPORTS MEDICINE DOCTOR

Sports medicine doctors specialize in treating athletes. They are skilled in all medicine, but focus particularly on bone, joint, and muscle health.

Sports medicine doctors often work in hospitals and clinics. They also work for professional or college sports teams. Some may attend sporting events in case of injury during a game.

Duties: Assessing injuries and other disorders in athletes, providing rehabilitation programs and care

Education: A qualification as Doctor of Medicine or Doctor of Osteopathic Medicine

Interests: Sports, helping others, problem solving

UTILIZATION REVIEW NURSE

Utilization review nurses are experienced nurses. They use their knowledge to determine treatment for patients, including those suffering from disease.

Utilization review nurses calculate the most cost-effective means of treatment and decide in which type of medical facility a patient would be most effectively treated.

Duties: Recommending the best types of treatments for patients in cooperation with other health professionals

Education: A qualification as a Registered Nurse plus two years' experience

Interests: Communication with others, problem solving, decision making

Activity: Medical Debates

It is increasingly possible to fight diseases and disorders by using genetic engineering. However, many issues related to genetics raise issues about altering inherited traits in humans. Choose one of the questions below and use it to prepare for a classroom debate.

DEBATE QUESTION: "Should people undergo genetic testing to discover if their children are likely to suffer from genetic disorders?"

Position 1: Yes—if they discover this in time, they can avoid having children or adopt children instead. This will prevent them from passing the disorder on.

Position 2: No—even if they have a disorder, people may still want to have children. Many people born with genetic disorders have full and rewarding lives.

DEBATE QUESTION: "Scientists have researched using genome editing to help the body fight diseases such as HIV/AIDS. Should they use this new technology in humans?"

Position 1: Yes—when fighting serious diseases that have killed millions of people, doctors should take advantage of all the methods available.

Position 2: No—we do not fully understand genome editing, and artificially changing people's genes may have unexpected effects. It might even help diseases to spread.

DEBATE QUESTION: "Should scientists be allowed to edit the genes of an embryo before it is born to make sure it has no genetic disorders?"

Position 1: Yes—genetic disorders can make people's lives very difficult, so anything that can prevent that possibility should be allowed, as long as it is carefully regulated.

Position 2: No—no one has the right to decide what sort of genes a child should have. What if the next step is to edit any embryos that are not seen as being perfect?

Test Your Knowledge

ONE
What is a vector?

TWO
What type of disease is passed on from parents to their children?

THREE
What are kidney stones formed from?

FOUR
Which parts of the body are affected by asthma?

FIVE
Gout is a type of what disease?

SIX
How does hemophilia affect the blood?

SEVEN
What is another name for indigestion?

EIGHT
What part of sunlight causes sunburn?

TEN
How is chicken pox transmitted?

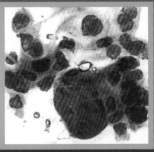

NINE
Who helped spread the germ theory of disease in the 1860s?

Key Words

abdominal: relating to the abdomen, the part of the body that holds the digestive organs

abnormal: different from normal

acute: describes a disease that begins suddenly or is of short duration

adolescence: the teenage years of growing up

allergies: disorders in which the immune system mistakenly recognizes a harmless substance as being harmful

arteries: vessels that carry blood from the heart around the body

arthritis: a disease that causes painful swelling in the joints

autoimmune: a disease in which the immune system attacks the body's own tissues

bronchi: air passages leading to the lungs

cancerous: related to the rapid and harmful growth of cells in the body

cartilage: a flexible tissue that connects bones

cells: microscopic units of which all living things are made

chronic: describes a disease that lasts a long time

clot: for a liquid to become thicker, like a solid

crystalline: formed of hard crystals

degenerative: steadily becoming worse

dehydration: a condition in which the body is short of water

diaphragm: a muscle at the bottom of the ribcage that helps with breathing

enzyme: a substance released by the body to achieve a particular result or effect

eradicated: wiped out

fecal: related to solid waste, or feces

genetic: inherited from one's parents

glands: organs that release chemical substances

heredity: the passing on of qualities from parents to their offspring

hormone: a chemical released in the body to control its functions

jaundice: a condition in which the skin and whites of the eyes become yellow

joints: points at which two bones are connected

nerves: long cells that transmit messages

palpitations: rapid or irregular heartbeats

pancreas: a large gland that releases enzymes into the stomach

parasites: organisms that depend on other organisms for their survival

pneumonia: a condition in which the lungs become inflamed

precautionary: describes an action taken to prevent an event from happening

protein: a type of chemical that helps build body cells and tissues

stools: pieces of feces

transfusions: replacing the body's blood with donated blood

ultraviolet: a type of light that cannot be seen with the naked eye

vaccine: a substance that is used to increase the body's resistance to disease

Growth, Development, and Reproduction of the Human Body

Index

LIGHTB☒X

✚ SUPPLEMENTARY RESOURCES

Click on the plus icon ✚ found in the bottom left corner of each spread to open additional teacher resources.

- Download and print the book's quizzes and activities
- Access curriculum correlations
- Explore additional web applications that enhance the Lightbox experience

LIGHTBOX DIGITAL TITLES
Packed full of integrated media

VIDEOS
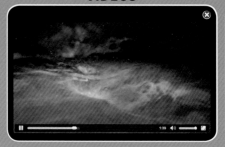

INTERACTIVE MAPS
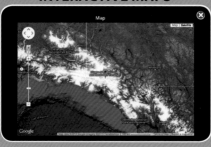

WEBLINKS

SLIDESHOWS

QUIZZES

OPTIMIZED FOR
✓ **TABLETS**
✓ **WHITEBOARDS**
✓ **COMPUTERS**
✓ **AND MUCH MORE!**

Published by Smartbook Media Inc.
14 Penn Plaza, 9th Floor
New York, NY 10122
Website: www.openlightbox.com

Copyright ©2021 Smartbook Media Inc.

Library of Congress Cataloging-in-Publication Data

Names: Shoals, James, author. | Willis, John, author.
Title: Diseases / James Shoals and John Willis.
Description: New York, NY : Smartbook Media Inc., [2021]

Series: Growth, development, and reproduction of the human body | Includes index. | Audience: Ages 12-15 | Audience: Grades 7-9
Identifiers: LCCN 2020004845 (print) | LCCN 2020004846 (ebook) | ISBN 9781510553873 (library binding) | ISBN 9781510553880 | ISBN 9781510553897
Subjects: LCSH: Growth--Juvenile literature. | Reproduction--Juvenile literature. | Developmental biology--Juvenile literature.
Classification: LCC QH511 .S496 2021 (print) | LCC QH511 (ebook) | DDC 612.6--dc23
LC record available at https://lccn.loc.gov/2020004845
LC ebook record available at https://lccn.loc.gov/2020004846

Printed in Guangzhou, China
1 2 3 4 5 6 7 8 9 0 24 23 22 21 20

052020
111619

Project Coordinator: Priyanka Das
Art Director: Terry Paulhus

Photo Credits
Every reasonable effort has been made to trace ownership and to obtain permission to reprint copyright material. The publisher would be pleased to have any errors or omissions brought to its attention so that they may be corrected in subsequent printings.

The publisher acknowledges Alamy, iStock, and Shutterstock as its primary image suppliers for this title.

First published by Mason Crest in 2019.